# ASTRO-THEOLOGY;

OR,

## THE RELIGION OF ASTRONOMY:

𝔣𝔬𝔲𝔯 𝔏𝔢𝔠𝔱𝔲𝔯𝔢𝔰,

IN REFERENCE TO THE CONTROVERSY ON THE

### "PLURALITY OF WORLDS,"

AS LATELY SUSTAINED BETWEEN SIR DAVID BREWSTER
AND AN ESSAYIST.

BY

## EDWARD HIGGINSON,

AUTHOR OF THE "SPIRIT OF THE BIBLE."

LONDON:

E. T. WHITFIELD, 178, STRAND.

1855.

TO THE

## Members of the Westgate Congregation, Wakefield.

---

I FEEL impelled by a sense of fitness, as well as that of duty and affection, to inscribe these Lectures to you, my Christian friends, to whom they were at first presented orally:—to some of you, as earnest and successful votaries of Natural Science; to many, as proficients in Literature and Art; to all, as Rational Christians, in whose free and comprehensive view of your Master's religion, the Gospel is seen to be in beautiful harmony with all the other good gifts of God to His creatures, and with the natural faculties, duties and hopes of His human offspring.

Such as has been, thus far, the unvarying spirit of my most healthy and happy connection with you as your Minister, may it be throughout !

WAKEFIELD, *Feb.* 1855.

# PREFACE.

It is my daily happiness, as one who holds a rational and free theology, to know and feel that Revealed Religion hails, instead of deprecating, the great discoveries of Science; in other words, that the voice of God in Creation harmonizes with that of His Gospel.

It is also, I grieve to say, my uniform observation, beyond the circle of avowed Unitarian Christians, that Science and Theology are accustomed to look upon each other with a greater or less degree of jealousy and suspicion;—the scientific man seldom daring freely to avow the most religious conclusions he draws from the study of Nature; the theologian dabbling very cautiously in the mere shallows of Science, lest he should plunge unawares into religious heresy.

The revival of the question of a Plurality of Worlds, in the hands of two such men as Sir D. Brewster and his anonymous, but well-known, opponent, has given occasion to both of them to confess the "religious difficulty" to which the passive, creed-bound theology of their respective churches subjects the man of science; while the one of those distinguished men boldly cites his orthodox theology in aid of his negative opinion as to the extent of a living Creation, and the other, holding a Plurality of Worlds too dogmatically for the free and candid spirit of Science, is obliged to strain his orthodoxy in a way that few, probably, will be found fully to approve.

I have read many notices of these two remarkable books of the day, in the periodical Magazines and Reviews, and have been grieved, rather than surprized, to find the same evasion of the scientific difficulty there, as in living society. Not only those organs of periodical literature which avowedly represent the orthodox theology, but also those many scientific and literary ones which dare not encounter theological suspicion, have cautiously left untouched the all-interesting question which these controvertists have painfully opened, How to reconcile Science and Revelation.

I am one of those who have been so happy as never to have imagined the possibility of their being at variance. Not profoundly scientific, I have ever revered, and to my limited ability and opportunity studied, Science as the expositor of God's own Works. Professedly devoted to Theology, I find its mighty truths accosting me in other records besides the Scriptures; and I find I can understand these venerable books best, and best appreciate their true character, in proportion as I bring an informed mind to their perusal. Most earnestly, therefore, do I desire to tell such of my fellow-students and fellow-christians as will listen to me, how musically those two divine voices sound, whether singly or united, if we but let them tell their own story, announce their own laws, sing their own poetry, and fill us with their own joy and love.

# CONTENTS.

When I consider Thy heavens, the work of Thy fingers,
The moon and the stars which Thou hast ordained;
What is Man, that Thou art mindful of him?
And the Son of Man, that Thou visitest him?
For Thou hast made him a little lower than the angels,
And hast crowned him with glory and honour.

PSALM viii. 3—5.

# LECTURE I.

## JEWISH ASTRO-THEOLOGY;

### OR,

#### HOW DEVOUT HEBREWS THOUGHT OF THE SUN, MOON AND STARS.

THE religious suggestions of the works of Nature are inexhaustible. Around, above, within us, God is everywhere proclaimed to the thoughtful soul. But perhaps no department of Natural Religion is more generally appreciated, than that which traces the perfections of God in the *heavens* above us. Things nearer and more connected with our daily secular use, are so familiarized and made common, it may be, that we too habitually regard them chiefly for their uses, and do not penetrate the mysteries and marvels which really lurk within the constitution and growth of all the living beings

that serve us, and the very corn and fruits that
bless us in the bounty of the year. The things
nearest of all to us, those of our own conscious
faculties and needs,—the traces of divine power,
wisdom and love in our own being, not merely
in our bodily frame, "curiously and wonder-
fully" as it is made, but in the spiritual faculties
of thought, feeling, sentiment, duty, worship,—
are among the later and more advanced studies
of the mind, when it has been already trained
to thought by reflection upon outward objects.
But the bright and majestic heavens, while present to every eye, yet removed by space and
mystery from the familiarity of rude thought or
low uses, have more constantly, perhaps, than
any other part of the Divine works, affected
mankind with a deep sense of that Divinity
which all things, when duly questioned, declare.
Not a child but wonderingly admires the golden
sun, the paler moon, the mysterious stars. Not
a neglected peasant's soul, not a heathen's natural thought of piety, but, when thinking of
Deity or aspiring to a future life, involuntarily
places the special abode of God and of good
spirits after death, among those " bright citadels
of light."

Nor perhaps can the philosophical mind, that ranges through the various realms of modern science, tracing time backwards through its geological ages till progressive Creation stands open before it in the gigantic folios of the earth's stratification, or reading those constant Laws which are recorded in the past as they are seen acting in the present, or enumerating the tribes, classes, species, varieties of each order of organized beings, and tracing their links of connection and analogy with one another,—whether prying into the profound secrets of the minute with the aid of the microscope, or into those of immense distance under the guidance of the telescope; — scarcely will the philosophical mind refuse to confirm the popular judgment of the peculiar *religiousness* of the impressions derived from the study, whether merely popular or more profound, of Astronomy.

The *vastness* of the heavenly bodies, when contemplated in the light of science, is so much increased to the perception of pure reason, above all that undisciplined imagination and adoring wonder had before ascribed to them, that the reflection becomes more and more profound, in which perhaps consists the peculiar religiousness of astronomical suggestions among those of natural religion in general,—the reflection, namely, How insignificant is Man, in the presence of such mighty works of God! And then that other reflection arises directly upon the traces of the former: How highly Man is endowed and privileged by his Maker! This is the order of the Hebrew Psalmist's thought:

" When I consider Thy heavens, the work of Thy fingers,
  The moon and the stars which Thou hast ordained;
  What is man, that Thou art mindful of him,
  And the son of man, that Thou visitest him!"

Thus far is the first reflection, under which man bows before the starry throne of Omnipotence, feeling himself the least of the Divine works. But the spirit that is in him rises in conscious dignity at the very thought which makes him thus feel his littleness. No other being on this earth can feel, can think, can know, how little or how great it is! The mighty Maker of all has given to man a distinguished place among those works, the mightiest of which is great and glorious only as reflecting His perfections. And from those works so vast and splendid, the mind of man turns to the nearer contemplation of others which display another and more engaging view of the Divine Maker; that is, as the Giver of Life, and, in life, of Happiness. All tribes of living creatures display, together with the wondrous power, the unbounded goodness, of the mighty Maker. And Man, placed at their head, lord over them by his intellect, and bearing the faint image, while he wields the deputed sovereignty, of God over them all, is now overwhelmed as much by the sense of the Divine favours thus heaped upon him, as he was humbled by the thought of the immensity of those works amid which he stands:

"For Thou hast made him a little lower than the angels,
　And hast crowned him with glory and honour.
　Thou madest him to have dominion over the works of
　　　Thy hand;
　Thou hast put all things under his feet:
　All sheep and oxen, yea and the beasts of the field,
　The fowl of the air, and the fish of the sea;
　And whatsoever passeth through the paths of the sea.
　O Jehovah, our Lord, how excellent is Thy name in all
　　　the earth!"　　　　　　　(Ps. viii. 5—9.)

The ancient Hebrews highly appreciated the religion of the heavens. Many of their sacred poems are, like the Psalm above quoted, written in express celebration of the attributes of God as seen there; and allusions to this class of thoughts dignify and adorn, in the very finest taste, their didactic and moral compositions. Distinguished by their noble belief in One Spiritual God, the Maker of all things, they more truly appreciated the beauty and grandeur of the Divine works than any other nations did, who, idolatrously worshiping the works themselves, missed the higher sentiment which those works are capable of exciting.

It is remarked of the Psalm before us, that the majesty of the heavens seems to be contemplated in it as seen by night, rather than by day. The moon and stars are specified, but not the sun. Perhaps this Psalm was first composed as an evening song of worship. How naturally may it have had its suggestion in the devout author's mind, in that calm period of nature's quiet and man's recovered peace, when the cares of the day are over, its work done, and if man goes forth into the field to meditate, be he king or be he peasant, with but an eye for beauty and a heart for worship (be he David keeping his father's flocks, or David on the throne of Israel), he sympathizes with nature's stillness, while the evening coolness refreshes and quickens his thoughts, tranquillized and concentrated as they are by the decline of the too glaring light; and while nearer objects around grow dim, the vault above grows more blue, and he sees the stars come forth and the moon walk in her recovered brightness, to the praise of Him who hath set His glory above the heavens.

There is another Psalm (the nineteenth), equally beautiful in poetry and pure in devotion, which is equally suitable, in its turn, to be regarded as king David's morning hymn of praise; when, risen early from sleep, and walking meditatively, perhaps, on his palace roof (the custom of the East being to make the roofs flat, and accessible for purposes both of retirement and of fresh air), he watched the dawn and then the break of day, saw in the roseate hues of the eastern horizon the "tabernacle set for the sun;" from which, "coming forth as a bridegroom out of his chamber, he rejoiceth, as a strong man, to run a race, going forth from end to end of the heavens, where nothing is hid from his heat." Then might the Psalmist have put this silent song of the morning into human thought and words:

One particular idea, and a striking phrase expressing it, is due to this high theology of the Jews.  They were struck, like other nations, with the regularity of the heavenly movements, and the stately magnificence of the marshalled gems of light above them.  They had no idea or rational conjecture as to what the stars might be; but they knew they were *the works of God.*  They were *as God had made them.*  They stood *as He appointed them.*  They moved *as He bade them.*  They waited upon *His sovereign word.*  They fulfilled *His almighty will.*  They were His Host,—His attendants, whether as the hosts of an earthly sovereign guard his person and swell his pomp in peace, or as they go forth with him to war.  Sun, moon and stars, are the host of the King of kings!  Mighty metaphor!  Great thought of a pure Theism, struggling for due representation!  What poetry so imaginative, while so chaste, as that of great religious thoughts?  This, then, was the religious philosophy of the Jews.  Of secondary causes and laws of nature, they understood no more than their neighbours; but of the final cause, the deciding and ruling Will, they knew more than philosophy has ever since added, in having learnt to believe in One Sovereign and Almighty Mind, above nature and its mighty and glorious works.

It must be a matter of surprise and regret to intelligent believers in revealed as well as natural religion, to find the authority of any name truly great in science attempting to uphold irrational expositions of Scripture, where Science and Scripture inevitably come into comparison, and may, by any but broad and comprehensive views of both, be easily set at seeming variance.

Sir David Brewster, in his part of the lately-revived controversy on the "Plurality of Worlds" (into which we shall look further by and by), has committed himself to the strange opinion that the Psalmist, whose words stand at the head of this discourse, was divinely inspired with astronomical knowledge unknown to his age, in order to write those fine words:

"When I consider Thy heavens, the work of Thy fingers,
  The moon and the stars which Thou hast ordained;
  What is man, that Thou art mindful of him?
  And the son of man, that Thou visitest him?"

He thinks this text "a positive argument for a plurality of worlds." He "cannot concur in the opinion of Dr. Chalmers, that a person wholly ignorant of the science of astronomy, and consequently to whom all the stars are but specks of light in the sky, not more important" (as he is pleased to add, though Dr. Chalmers did not say so) "than the *ignis fatuus* upon a marshy field, could express the surprise and deep emotion of the Hebrew poet." He therefore "cannot doubt that inspiration revealed to him the magnitude, the distances and the final cause of the glorious spheres which fixed his admiration." So then this veteran philosopher, who knows the whole history of astronomical discovery, and has observed how the progressive knowledge of the heavenly bodies has been laboriously gained through centuries of noble discipline alike to the intellect and the character of the sons of science, actually believes that the Hebrew Psalmist was inspired with a knowledge of the magnitudes, the distances and the final cause of those glorious spheres which fixed his admiration! How meanly must he think of the functions of an inspired astronomer, that this Psalmist never communicated such august knowledge to those around him! Natural science, gained by the unaided human faculty, delights to spread itself in free communion between mind and mind; but an *inspired astronomer*, we

are to understand, left the world in total ignorance of the scientific truths which had been
imparted to him! Is Inspiration, then, thus
niggard and jealous, while Science is so diffu-
sive and generous? O, shame upon those re-
presentations of Religion which make it chargeable with such unworthy churlishness!

But this philosopher (and a great physical
philosopher he is, though a narrow theologian)
discovers yet deeper insight to have belonged to
the Psalmist. He assumes also that the Hebrew
poet regarded man, not only (as the poet describes him) as "made a little lower than the
angels, and crowned with glory and honour,"
but also as a being "for whose redemption God
sent his Son to suffer and to die;" and then
(with the help of this strange anachronism of
thought) he argues that man, thus regarded,
"could not be an object of insignificance in the
Psalmist's estimation; and, measured therefore
by his high estimate of man, his idea of the
heavens, the moon and the stars, *must have been*
of the most transcendent kind. Had he been
ignorant of astronomy (our philosopher persists),
he *never could* have given utterance to the sentiment in the text." And so he concludes that

the Hebrew poet "doubtless viewed these worlds as teeming with life, physical and intellectual, as globes which may have required millions of years for their preparation, exhibiting new forms of being, new powers of mind, new conditions in the past, and new glories in the future."

So then, according to one of the most scientific men of our day, the Hebrew Psalmist was miraculously inspired to know all that sages have been learning through ages since, of the sizes and motions of the heavenly bodies and the geological structure and history of the earth, and all that they have been inferring from the analogies most accessible to us respecting the possible or probable condition of sun, moon and stars, as habitable worlds;—all this (he declares) must have been miraculously infused into the Psalmist's mind, when he gave utterance to his noble song of praise to God for His works!

Now here is a most serious damage done to the credibility of the Scriptures, if it does not also excite a suspicion against the very exalted scientific views of the writer. The true explanation seems to be, that Sir D. Brewster (who is now in his 73rd year) was in early life educated for the ministry of the Kirk of Scotland;

and though he soon gave up the profession of theology for that of science, his early theological impressions appear to have retained their strict accordance with the rigid orthodoxy of Scotland. In the discussion in which he is engaged with an anonymous, but pretty well ascertained, author (also high in the scientific world, if report correctly guesses his name, and high in university honours and preferments as a divine of the Church of England), on this question of a Plurality of Worlds,—or, in other words, the question whether any of the heavenly bodies may reasonably be believed to be inhabited,—it is curious and sadly instructive to notice how "religious difficulties," confessedly arising out of their mutual orthodoxy, seriously perplex the scientific question on both sides, causing great annoyance and tempting to special-pleading of a most unphilosophical kind on the part of Sir David, and giving room for the suspicion that his opponent has been not a little influenced in his rejection of the plurality of worlds, by feeling such a belief to be incompatible with his orthodox theology. Of this I shall speak more pointedly in its proper place, when we pursue the progress of these great astro-theological ideas, and endeavour to realize to our religious thought the general evidence on which these curious and interesting, yet abstruse and confessedly doubtful, speculations hang.

For the present, let us return for a few moments to the Hebrew astro-theology, without regard to Sir D. Brewster's imaginative additions,—not asking nor conjecturing what the Psalmist *must* have thought of the stars when he exclaimed, "What is man?" under the prior supposition that he knew how Christ was to die for man,—but remarking simply what he and other pious Hebrews really have expressed themselves as thinking and feeling in reference to the works of God before their eyes. Those thoughts they have expressed at least to the following effect.

I. To the unscientific, yet intelligent and religious-minded beholder of the hosts of heaven (like king David and many other Hebrew psalmists and prophets), the daily and nightly sky appeared as a scene at once of *beauty and grandeur*, impressing them with sentiments of *religious delight and love* towards Him whose hand made all these things.

It is so with us before we learn our first

lessons in scientific astronomy,—before we know whether the heavens move daily round to shew themselves to us, or the earth revolves to let us look at every part of them in turn. And when we have read the history of astronomical thought and discovery, and learnt from Copernicus and Newton the indisputable order of the Solar System, while this sublimest of human knowledge excites our deeper and maturer reverence, the bright and beautiful phenomena themselves still offer to our senses the same impressions as at first, which the associated thought, however, renders proportionately more beautiful and grand to our appreciation. Their beauty and magnificence touch *our* sense of religion, just as *theirs* of olden time, because we see them shining not by their own power to their own praise, but to the praise of Him who "hung those lamps on high." Ancient Jewish and modern Christian sentiment coincide here:

"These are Thy glorious works, Parent of good,
　Almighty! Thine this universal frame
　Thus wondrous fair! Thyself how wondrous then,
　Unspeakable! who sitt'st above these heavens,
　To us invisible, or dimly seen
　In these Thy lowliest works; yet these declare
　*Thy goodness* beyond thought, and *power divine!*"

II. The old Hebrew worshiper was devoutly impressed, much as the modern Christian philosopher is, with the *order* and *permanency* of the Divine works in the heavenly firmament. Order was their believed *moral* law, the law of evidently presiding Will and Purpose, long before the *physical* laws to which the preservation of their order may be in part at least intrusted, were known or thought of. Observation proved it as a fact, long before science was able to account for it. "For ever, O Jehovah," said Hebrew piety, reading by the soul's insight the golden hieroglyphics of the sky,—"For ever, O Jehovah, Thy word is written in the heavens!" "They continue unto this day according to Thine ordinances, for all are Thy servants!" And this perception of order among things so vast and distant, so numberless and so mysterious, naturally became a deep sense of *reverence*. Without any theory or conjecture, as far as we know, respecting the nature of the heavenly bodies,—without the slightest idea (that we can be justified in ascribing to them) approaching to the modern scientific thought of a plurality of habitable worlds,—the old Hebrews read, in the orderly phenomena of the starry heavens, true lessons of reverential piety. Hear the author of the book of Job expound this moral, as in the name of the Most High himself speaking to mortal man. What is man, indeed, in the presence of these mighty and seemingly everlasting revolutions?

> "Canst thou bind the sweet influences of the Pleiades?
> Or loose the bands of Orion?
> Canst thou bring forth Mazzaroth in his season?
> Or canst thou guide Arcturus with his sons?
> Knowest thou the ordinances of heaven?
> Canst thou set the dominion thereof in the earth?"

When the devout philosopher of modern times contemplates this sublime order of the heavenly bodies as the result of divinely-appointed Laws, the real image still effectually before his thought is not that of the material law which instrumentally produces it, but of the ever-recurring phenomena which are thus produced, the periodical rounds which he has himself seen again and again exactly fulfilled, and which he believes to have been fulfilled with the same exactness through all the ages that have intervened, since the pen of that old Hebrew poet recorded in those fervent words his reverential thought. If he penetrates through the phenomena, it is not to rest upon a material law, but to rise to the central Will. Science endorses all that the Hebrew poets have sung in praise of the majestic uniformity of Creation.

III. And again: with this lively sense of the beauty and greatness of the starry works of God, and this profound reverence for their majestic order, the Hebrew worshipers united another active sentiment, which it becomes us to feel also; namely, that of *thankfulness* for the evident subservience of these mighty works of God to the *comfort and happiness* of human life. The most unscientific knows that the sun is designed to shine by day, and the moon and stars to shine by night. He believes they are made for the purpose (as he sees them serve that purpose) of marking out times and seasons and days and years. And in words to this effect the Hebrews of old spoke reverently and thankfully of the blessings conferred upon man's abode by the bright host of heaven.

All such words have a deeper scientific truth than those who first spoke them knew. The growing science of Astronomy has shewn how the celestial phenomena not only mark, but *make,* our times and seasons, and confer their bounties duly as they roll; while it forbids us to regard many of those uses as limited to the Earth on which we dwell, and lifts our adoring thought to other globes, like ours, yet different, which receive corresponding yet varied influences, as parts of the same great system of Worlds. And, with deep reverence and devotion, Science asks us whether we can doubt that in other worlds besides our own,—how many or how few besides, we may not dare to guess,—there are sentient creatures to receive their Creator's overflowing blessings, and intelligent beings to see and love and worship Him in His gifts.

# LECTURE II.

SCIENTIFIC ASTRO-THEOLOGY;

OR,

MODERN PHILOSOPHICAL VIEWS OF THE SOLAR SYSTEM
AND FIXED STARS.

---

WE have endeavoured to realize the religious
lessons of the heavenly bodies, as they shone
before the eyes of the devout Hebrews of old;
who, long before the times of Astronomical
science, looked upon all things as the work of
a Supreme Creator, with sentiments of delight
and love for the beauty and grandeur that were
around them, with profound reverence for the
order and permanency of the mighty and mysterious heavens, and with devout gratitude for
the beneficent uses fulfilled by them in the
changes of day and night and the revolutions
of the seasons, with which they were evidently
connected.

It needed no scientific ear to hear the heavens telling the glory of God. It required no telescopic vision to see His handy-work as shewn in the firmament. A deeper insight into the religious import of Creation had been given to these Hebrews by their sublime doctrine of One Almighty Maker and Lord of all things, than to the early astronomical observers by their very imperfect means of observation. The profoundest sense of the Divine majesty and bounty, and of man's littleness amid the creation, yet of the near place he occupies in the Divine love and care, dwells in those words in which the Psalmist points the contrast and cherishes the emotion due to the contemplation of the bright heavens above his head: "When I consider the heavens," &c.

Yet, with the advance of Astronomical science, the religious mind has found ever new sources of wonder and devotion. Thought has been strained to its utmost, to grasp the mere facts, as ascertained and expressed in number, size and distance. Imagination has confessed itself unable to realize the suggestions which crowd upon it irresistibly from those facts. Devotion has bowed lower, and felt what Sacred

Mystery means, in things perplexing not by their contradiction but their vastness, when, through the gradual ascent of knowledge, man has found his intelligent conviction leaving intelligent perception far behind, and the noblest efforts of his finite faculties have been baffled by the Infinite that is disclosed to him. O, not in any subtleties of metaphysical thought, not in the daring contradictions of dogmatic creeds, not in any unnatural (or non-natural) meanings ascribed to customary words, is the real idea of Sacred Mystery to be found; but in the Immensity which baffles our conceptions,—in the Complication which, orderly as it is, eludes our effort to trace it through its innumerable ramifications,—in the Secrecy in which causation still hides itself beside the throne of God !

The mysteries of Creation are its marvels still untraced, or only traceable until, sooner or later, we are lost as we approach their infinite source.

The mysteries of Providence are the secret purposes, so far beyond our reach to unravel, for which all the joys and trials, all the temptations and all the discipline of mortal life is sent.

And the mysteries of Scripture are still true to the same sense, as the secrets of the Supreme

Mind; for the Scripture shews us many mysteries *revealed*, in its successive disclosures of the
Divine Will and human duty and hope; while
not even Revelation has revealed *all* the secrets
of life, death and immortality, because to *finite*
man such a revelation would be constitutionally
impossible.

In fact, close on the borders of our knowledge,
whether scientific or directly religious, and whether natural or supernatural, the dark shadows
of the unknown still rest, and still would rest,
however far those borders might be extended;
or else the dazzling haze of immensity not less
effectually defies our mental vision. Such is the
real mystery of God, and of His works and ways!

The progress of astronomical discovery has
developed this true sense of the mystery of
Creation. It has given vastness, before inconceivable, to those bright specks that once seemed
but to adorn the blue arch of night; and, from
the known vastness, it has plunged the mind
into immensity beyond both knowledge and
conception. But it has, meanwhile, irresistibly
prompted the inquiry into the possible or probable *uses* of the heavenly bodies, and has suggested to us the grandest thought that there is

in the whole range of Natural Science and Religion, namely, that of a *Plurality of Worlds*.

While this fair Earth upon which we dwell was regarded by its rational inhabitant as the centre of creation,—while its hundreds and thousands of miles, so vast to his experience and reaching far beyond his farthest travels, was practically his measure of conceivable size and distance for everything else,—while he believed that the sun revolved daily from east to west for the earth's sole benefit; and, without the slightest knowledge, thought or belief as to the sizes or distances of sun, moon or stars, was satisfied to regard them as ministering to the beauty, comfort and convenience of man's abode, —so long the human race was, in the eye of human reason and religion, the *ultimate purpose* of the universe. Man, the head of creation on this earth, naturally believed himself the sole created intelligence among God's works; or if, under the inspiration of high religious thought, he believed (as the Hebrews did) in the existence of spiritual or angelic agents higher even than himself, they were not regarded as the inhabitants of the heavenly *orbs*, but rather of the mysterious heavenly *arch* in which those orbs themselves stood as the attendant host, and their office was supposed to be that of administering the providence of God to His creatures on this earth.

But after the discovery that this world, instead of being the centre of the whole system of things (and that system small, proportionately to that idea), is only one of many planets which revolve, like it, around the sun, illustrating the same great and uniform laws of matter and motion, heat and light, and sharing a similar alternation, varied in detail, of day and night and also of seasonal changes,—it was impossible for religious minds to resist the suggestion of analogical thought whispering to them, If, by these arrangements, this earth is fitted for the residence of man and all the other living and rejoicing creatures that dwell upon it, is it not probable that those similar, but varied, arrangements in other planetary bodies, may subserve corresponding uses for various other orders of living and happy creatures, with intelligent beings at their head also?

The argument from analogy put itself into this form, long before those minuter observations could be made upon the various planetary bodies,

by which the scientific astronomer has been en-
abled more recently to tell with considerable
confidence, at least in many cases, how far their
physical circumstances correspond to, or differ
from, those of our globe. And it is very possible to generalize somewhat too fast and to infer
too confidently, till this truly grand, sublime
and religious idea should verge upon the impro-
bable or the absurd. But, on the other hand, the
discriminating circumstances revealed through
the recent wonderful improvements in the telescope, if they make it more difficult to conceive
that certain globes in our system are inhabited
by forms of life at all resembling those with
which we are familiar, render it more difficult
still to doubt it as regards those others which
are proved to resemble our globe very nearly
in the physical circumstances which specifically
adapt it for human residence.

Then another thought, yet more vast, and
utterly bewildering from its vastness and from
its vagueness, has followed in the astro-theology
of later times. From the knowledge that our
Sun is the centre of a great system of planetary
worlds, it is inferred, with as high probability as
the nature of the case admits, that the Fixed

Stars, those specks of light which were but the lamps of the sky to earlier ages, may be so many Suns, like our own in purpose, giving light and heat to planetary systems corresponding to ours, and that some or all the planets which incircle them may be worlds full of life and happiness, as this Earth is.

Reason shrinks indeed, and imagination itself totters, under the mere attempt to realize the facts and but to fancy the suggested inferences. Around our Sun, this Earth revolves at a distance of nearly one hundred millions of miles ;— and it would be well to pause a moment and just ask ourselves, whether we *really apprehend* what one million is. A million is as easily said as a hundred or a thousand; but a hundred must be counted ten times over to make a thousand, and that thousand must be a thousand times repeated to make the one million; and then that million must be counted through its thousand-thousands yet ninety-five times over, to make the miles that reach from the Earth to the Sun. At such a distance the Earth revolves round the Sun, travelling above a thousand miles every minute, the attendant Moon revolving monthly round it in its course. Within this annual circle

described by our Earth are two other planets,
Venus and Mercury, and without it another,
Mars, about half as far again from the Sun,—
the two that are nearest to us having a very
close similarity to our globe in all those things
which specifically fit this world for its specific
inhabitants.   Beyond, at five times our distance
from the Sun, is the magnificent planet Jupiter;
at nine times our distance, the mysterious Saturn,
girt with his triple ring of light; at eighteen
times our distance, there is Uranus; and at
twenty-eight times our distance, is Neptune;
all these encompassed with Moons,—four, six,
eight in number,—carrying out (as if with some
view to compensation by increase of number)
the analogy of our Earth and its Moon, while
other points of practical resemblance diminish
rapidly by the mere law of distance.   These,
with a multitude of small planetary bodies situ-
ated between Mars and Jupiter, and some more
mysterious bodies still, of unknown constitution
and use, the Comets, form our Solar System.
And then we are taught that the Fixed Stars
are, to all appearance, Suns like ours, shining
by unborrowed light,—all of them so distant,
that none of them seems in the slightest degree

nearer or more remote, for all the 190 millions of miles that we change our place with respect to them in six months' time; — and, at such distances, how immense must they be in size! Then, for their number! Every improved telescope, looking further into space, has seen them only more numerous than before, and still without any intimation of approaching limit! No bounds to space, and no space unstrewed with suns! And these thousands upon thousands of suns, we can scarcely avoid concluding, are (or it is more reverent, perhaps, to say, may be) each encompassed by a system of worlds, more or less analogous to that to which our Earth belongs! Who can realize the idea? And then, to attempt in thought to give yet further to all those planetary worlds their myriads of living inhabitants! It is too much! too much! And yet, who shall deliberately say that less than this is justified by appearances?

It is, however, quite possible that the zealous advocate of the Plurality of Worlds may overstate his case, presuming to assert more than analogy fairly justifies, and by so doing may excite doubt instead of reverence.

Looking more carefully at the Solar System,

D

with all the wonderful aids of modern science
and art, the philosophical astro-theologian is far
from maintaining that *all* the bodies which form
that system must be, without exception, habitable worlds.   The three bodies within, and the
one next without, the Earth's orbit,—Mercury,
Venus and Mars, — are those which bear the
closest analogy in all respects to the Earth ;
their periods of day and night being nearly the
same as ours, and the inclination of the axis
of Mars (the cause of seasons) being almost the
same as that of the Earth; while all of them
have atmospheres, and in all of them the force
of gravitation, and the amount of light and heat
communicated from the Sun (without reference
to unknown causes of modification which may
operate), are such as imagination can reconcile
even to the condition of beings similar to those
which dwell upon our own planet.   But the
planets more distant from us in position, differ
much more widely in all those important conditions which depend upon distance from the
Sun, and in many other conditions also; and
this consideration (on the very principle of analogical reasoning, which should notice differences
as carefully as resemblances) makes it necessary

that we should speak less confidently, at any rate, as to the *kind* of beings for whose residence we venture to suppose those planets may possibly be adapted, and hold our imagination more free from all the actual forms of being which belong specifically to this Earth. As we could not have conceived of the existence of creatures breathing in the water, if we had seen only those that breathe the air; so we ought neither to deny the existence, nor to attempt to define the organization, of creatures that may be fitted for planetary abodes where we and the other creatures of this Earth could not live. Then, the Sun itself, the centre of motion to the system,—though it is perfectly conceivable that its surface, beneath a radiant atmosphere, may be the abode of suitable life,—yet this body might, if any difficulty be felt on this head, be regarded as uninhabited, without at all impugning the general doctrine of the Plurality of Worlds. A sufficient *use* in the purposes of the system would have been assigned to the Sun, as the centre of motion and the source of light and heat to the rest, though it be not itself regarded as an inhabitable world. The Moon, again, the comparative nearness of which to the Earth renders the observations made upon it in this respect quite reliable, is found to be destitute of an atmosphere, which the primary planets, generally at least, appear to have; and this, together with other considerations, may seem decisive against the idea of the Moon being inhabited, yet still without impugning the general doctrine.   The Moon is the attendant of the Earth; and her functions so important to the Earth in the creation of its ocean tides (and aerial tides also, with great and beneficial influence, no doubt, upon the weather), besides her important use to us as a luminary by night,—these functions are quite sufficient to vindicate creative wisdom and goodness to our most scrutinizing thought, without our endeavouring to believe, against appearances, that the Moon too has its inhabitants.   And the same train of reasoning makes it perfectly unnecessary for us to endeavour to assign inhabitants to the Satellites of Jupiter, Saturn, Uranus and Neptune.  Again, the Planetoids, as they are called,—the thirty or thirty-one little planets (if another or two be not added to the discovered number while we speak about them),—those little planets which lie next beyond the orbit of Mars, are so peculiar and exceptional a group among the bodies which circulate round the Sun, that the argument from analogy, instead of peopling them, would hold the question of their habitableness in doubt at the very least. They are, we may say, close together, thirty or more small bodies, some scarcely larger than a mountain, circling round the Sun while crossing each other's orbits, and irresistibly suggesting to the Astronomer's mind the idea either of a world destroyed, or of planetary matter destined possibly to become some time a world! Nor, again, does this belief in a Plurality of Worlds make it necessary that we should believe the Comets to be inhabited. We know far too little about their nature, to believe confidently that they are; and we know many things respecting them which remove them so far from all resemblance to the known habitable world, as to make it seem probable that they are not. While they are subject to the same great laws which guide all the planets round the Sun in elliptical orbits, the ellipse is so slight in all the rest, and so exaggerated in the comets, necessarily involving such extremes of heat and cold, while their substance appears to be so attenuated, that analogy would dispose us to doubt or deny, rather than to affirm, their habitableness, and to keep in cautious suspense our judgment of the purposes which they may serve in the System.

He then who, to complete a theory of the Plurality of Worlds, dogmatically asserts that all these must be inhabited, invites (and indeed receives) the ridicule of the scientific opponent; who, in his turn, artfully begins the survey with these weakest parts of the case, and after scornfully asking whether the comets and planetoids, and even the nebulæ, can be rationally considered to be the orderly abodes of intelligent beings like man, comes last in order to the worlds most like this, and then doubts the theory just where he might (had he begun there) have believed it probable and felt it to be devout.

And when the same analogy is extended, with such hesitancy as becomes us, to the possible systems of worlds which we suppose may incircle the Fixed Stars,—it does not suggest to us to regard those myriad suns as themselves habitable worlds, so directly as to believe it possible, yea probable, that they may have revolving round them planets analogous to those of our

Solar System, which may receive light and heat and blessing from them, as we do from our Sun.

Such is the outline of the analogical argument from the well-known parts of the Solar System, to the less known, and from the whole, to other Systems in the starry heavens. There should be no temptation to overload or exaggerate this vast argument. At the least, it carries our minds quite beyond all clear conception of size, distance and numbers, in what we are forced to admit at the mere dictate of scientific astronomy, which has actually measured planets and their spaces by the application of the same optical and mathematical principles by which a survey of our own mountains and plains is effected, without contact with every part.

Few men at once scientific and religious, probably, doubt the great inference suggested, of the Plurality of Worlds. It has seemed to be the involuntary conclusion of ever - widening science.

But Religion (or rather, what was thought to be Religion) hesitated long to admit the true theory of the universe, in opposition to the senses of the ignorant, which seemed to tell them that the Earth does stand still, and that

Sun, Moon and Stars do move round it,—and in opposition also to the supposed dictate of the Scriptures, which when they speak (as even a scientific man may now speak) of the Sun as *going forth* and *rising* and *setting*, were stupidly supposed to utter oracles of scientific truth, and were appealed to by the Church and the Inquisition against the genius of Copernicus and Kepler, and even against the telescope of Galileo.

Strange indeed have been the alternations of scientific belief as to the constitution of the Solar System. Pythagoras, five centuries after the days of David and five before the Christian era, acutely guessed the actual order of things, mingling his noble theory with strange but beautiful fancies of geometrical harmonies and music of the spheres. But all this was, by the great astronomer of the second century after Christ, Ptolemy, completely swept away from scientific thought, as though it had never been; the Earth was restored to its popular place as the centre of the universe, and Sun, Moon and Stars were bidden to revolve again about it. It was not till above thirteen centuries more had passed away, that Copernicus, himself a priest, reproduced the thought, and, amid opposition calling itself religious, again presented the beautiful, simple and majestic truth to the delighted minds of those who were capable of appreciating it and dared to do so.  But the religious prejudices of his time still forbade its general reception; and towards the end of the sixteenth century, Tycho Brahe ingeniously, yet one can hardly think sincerely, endeavoured to conciliate religious prejudice by a compromise with scientific truth, permitting Mercury and Venus to revolve round the Sun, but insisting that all the three should, with the more distant planets, revolve daily round the Earth as the centre of the whole system.  But Kepler followed hard upon Tycho; and amid priestly interference prohibiting the publication of his books and placing them, with those of Copernicus, in the *Index Purgatorius* of the Church, he established his three great Laws of planetary motion; on his publication of the last and grandest of which, he, with true sublimity, unsurpassed in the devout annals of Science, writes thus to a friend: " The book is written;—to be read either now or by posterity; I care not which. It may well wait a century for a reader, as God has waited six thousand years for an observer." And Galileo was Kepler's contemporary, and his correspondent; on whose invention of the telescope, and its disclosure of the four moons of Jupiter moving regularly round their primary, the truth of the Pythagorean or Copernican system was demonstrated to men's sight, and (the Inquisition notwithstanding) it was made manifest that the Earth did move, and that Mind was moving too;—that Society was stirred, Science on the advance, Religion reforming fast. And would that Science had thenceforth always known and asserted her own pure religiousness; and that the expounders of Scripture had known a truer reverence for the Works, and a more discriminating one for the Word, of God; so that those perpetual jealousies of the Church against Scientific truth might not have again revived,—as, alas! they have done so often,—disputing the truth of God's own Works, and bringing that of His Revelation also into doubt by the ignorant jealousy of its guardians and expounders!

Yet Science cannot give up her facts, though timid advocates may tamper with them. And happy indeed are they who have so read the revelation of Divine truth in the Bible, as to

catch the devotion of the Hebrew Psalmists and
Prophets without consecrating the philosophical
errors of their day, and to read the Creation
with their souls awake, without ever suspecting
that they can thereby dishonour Christianity.

*Chronological Note to the foregoing Lecture.*

| | |
|---|---:|
| Pythagoras born . . . . . . . . . B.C. | 590 |
| Ptolemy flourished about . . . . . A.D. | 140 |
| Copernicus born . . . . . . . | 1473 |
| Tycho Brahe born . . . . . . . | 1546 |
| Kepler born . . . . . . . . . | 1571 |
| Galileo born . . . . . . . . . | 1564 |
| Galileo invented telescope . . . . | 1609 |
| Galileo forced by Inquisition to abjure his science . . . . . . . . . | 1633 |
| Des Cartes born . . . . . . . . | 1596 |
| Newton born . . . . . . . . . | 1642 |
| Fontinelle (author of "Plurality of Worlds") born . . . . . . . . | 1657 |
| Derham (author of "Astro-theology") born . . . . . . . . . . | 1657 |

# LECTURE III.

## ORTHODOXY AT ISSUE WITH THE CREATION;

### OR,

### THE "RELIGIOUS DIFFICULTY" CONFESSED BY SIR DAVID BREWSTER AND HIS OPPONENT.

---

WE have learnt, from one of their Psalmists, how the pious Hebrews of old interpreted the religion of the sky,—possessed as they were of little scientific knowledge, unable to conceive the sizes and distances of the heavenly bodies, and of course intirely ignorant of the laws by which their motions are guided; yet with eyes to discern their beauty, with admiring perception of their order and constancy, and thankful recognition of their evident benefits to mankind; and, above all, possessing that sublime theology which ascribed Creation and Providence to One Supreme Mind, and led them to look upon all the starry host as the ministering servants of the Most High God. Those words of the Hebrew Psalmist express therefore the essential thought of the devout, but unscientific, observer, in ages also since his own.

We have carried the thought with us (in the second Lecture) along the path of scientific astronomy; and have found the noblest and truest commentary upon it, not among Churchmen, who have too often made the letter of Scripture untrue by exaggeration or irrational interpretation, but among Mathematicians and Sages, realizing its spirit continually more and more in the widened pages of the sky; in the wonderful insight of Pythagoras, the scientific devotedness of Copernicus, the mathematical genius of Kepler, the sight-convincing telescope of Galileo, the comprehensive and all-harmonizing theory of Newton. These great enlargers of our knowledge have felt that, as their knowledge of the universe advanced, whether in exactness or in extent, it furnished the more solid basis continually for those devout conclusions which, in the minds of the Hebrews, had anticipated the course of scientific astronomy. But—O, the injustice, alike to those bright Hebrew visions

of God amid His works, and to the patient
explorers of the Divine works ever since!—we
have found the scriptural theologian, at every
step, the most bigoted opponent of scientific
research, by his narrow idea of a verbal inspiration investing mere popular ignorance with
divine authority. We have seen the Church
anathematizing the movements of the Solar
System, and the Inquisition visiting them with
pains and penalties!

These religious difficulties lying in the way
of scientific conclusions, were caused—as similar
ones still are caused—by mere *ignorance* of the
facts of science allying itself to the deeply reve-
renced, but thoroughly misunderstood, *letter* of
the Scripture. But there is another kind of
religious difficulty which now sorely perplexes
the astro-theology of the Christian world, arising directly out of its metaphysical *creeds and
theories*, quite extraneous though these really
are to the Scriptures, yet of co-ordinate authority in most minds. Through a process of reasoning as ingeniously subtle as it seems perverse
to those who do not accept it as true, the " orthodox" believer (so called) who has surmounted
the vulgar idea that he is forbidden by Scripture

to believe in the motion of the earth, feels himself on the verge of the most awful religious
heresy when he admits the great scientific idea
of a Plurality of Worlds! This "religious
difficulty" (as it is called by one such orthodox
believer), this "astronomical objection to religion" (as another calls it), is confessed on the
one hand and urged on the other, by the two
great scientific disputants who have lately revived
that most interesting and ennobling question of
scientific religion, the Plurality of Worlds.

I intreat you to view the matter just now in
the aspect in which these men think it necessary,
as orthodox theologians, to present it, that you
may see in what peril Religion itself is placed
by the popular theology, according to the avowal
of orthodox men themselves.

To those to whom natural religion and revealed are but as harmonious voices hymning
the praises of the One Supreme Beneficence,
this perplexity appears indeed to be the most
perverse and gratuitous of religious difficulties.
But to the bulk of those sincere believers in the
popular form of Christianity, who have reasoning
power enough to trace the connected thoughts,
it is a real and most alarming difficulty,—it is

a rock that threatens their Christian faith with shipwreck.

The difficulty is precisely this:

According to the Trinitarian belief, in all its many modifications, the death of Jesus Christ, as the Second Person of a supposed Divine Trinity, was the means of procuring salvation for the human race, or for a certain part of them, who must otherwise have been eternally lost from their Maker's blessing.

Now, on acquiring the sublime views, which Astronomy opens to us, of other worlds more or less resembling our own, and naturally inferred to be (or some of them at least) inhabited by intelligent beings more or less analogous to our own race,—the "orthodox" believer inevitably asks himself, what he is to think of the salvability of those other intelligent beings in other worlds. He cannot simply leave this new glimpse of other habitable worlds to the providential care of Him whose inexhaustible beneficence they seem to proclaim. The Theist of Nature can do that. The Unitarian Christian can do it; and feel the happier for even the vaguest thought thus gained of the diffusiveness of Divine love. But the orthodox Christian cannot. So,

at least, it is declared by these his representa-
tives in the world of Philosophy. His theological system bids him ask, and answer to his
own creed's satisfaction, these questions:

Do the intelligent beings who may be supposed to people other worlds, need a Redeemer,
to do for them what Christ did for the inhabitants of this world? Orthodox believers generally (but not universally*) conclude that they
do.

If so, How do they find that Redeemer? is
the next question.

Can the Second Person of the Divine Trinity
be believed to have made his incarnation and his
expiatory sacrifice in each habitable world in
succession? The orthodox believer, it would
seem, cannot admit this thought for a moment.
I know not exactly why; but he cannot, he
does not.

Can the expiatory sacrifice of Christ then,
presented on this earth for mankind 1800 years
ago, be conceived to have had a similar influence

* The late Dr. Chalmers thought it possible they might
not; and he is deemed evasive, if not heretical, by Sir D.
Brewster, for having admitted such a possibility. *Rational*
beings, it is thought, *must* have fallen everywhere.

E

throughout all the habitable worlds of God's universe? The orthodox believer, who admits a plurality of worlds, is generally driven to this conclusion!

Yet how hard it is for a really scientific man to make this a part of his belief! When, in the wider atmosphere of Science, he has learnt to look at this fair globe on which we dwell, not as the centre of the Universe, but as one small part even of the Solar System, and that System itself as but a small part indeed of the boundless Universe,—it seems like undoing his scientific thought, to come back in the name of his theology to this same Earth as the centre of the most stupendous spiritual influences to the whole creation. This little Earth is arbitrarily selected from among all the planets of all the Solar Systems in the Universe, to become the centre of a Spiritual Force, before which the functions of gravitation itself in the material creation appear insignificant! Such is the orthodox Christian's astro-theology. It is difficult for him to realize to his own belief his own theory, and solemnly to say to himself that he thinks the inhabitants of Venus and Mars were redeemed from sin by Christ's dying at Jerusalem; and not only so, but that all the habitable planets which he conceives of as probably circling round the Fixed Stars, also recognize this planet Earth (invisible to them, as those planets are to us) as the centre of religious influences the most mysterious but most potent, to them and to the universe. One would almost think such difficulties were strained and imaginary. But no; they are indeed real to the sincere believer in what are called orthodox views of religion, for they are the legitimate consequences of his belief. Astronomy furnishes perhaps the severest scientific test of our Platonic and Middle-age theologies. It brings them face to face with the mighty Universe, and in its wide field reduces them at once to their due proportion in the possible or probable spiritual world, when the Earth itself is seen taking its subordinate place in the stupendous march of the Creation. And then the orthodox Atonement is proved to be a paralogism in astronomy, as palpably as the Athanasian Trinity is an absurdity in arithmetic.

Hear the learned and scientific, and both of them, alas! orthodox, disputants in the matter before us, severally confessing the difficulty which I have endeavoured to describe.

The author of the first book, intitled, "Of the Plurality of Worlds, an Essay," who takes the negative side of the argument (and whom we must scruple to speak of as Dr. Whewell, the Master of Trinity College, Cambridge, only because his name is not actually on the title-page of a book universally ascribed to him and not disowned by him), supposes "a bold advocate of the Plurality of Worlds" to say:

"The only matter which perplexes us, holding this belief on astronomical grounds, is that we do not quite see how to put our theology into due place and form in our system."—P. 216.

Again he describes the difficulty as taking this shape:

"If we believe the astronomers, will not such a belief lead us to doubt the truth of the great scheme of Christianity, which makes the Earth the scene of a special dispensation?"—P. 129.

In the Preface to his second edition, he makes one of the speakers in an imaginary dialogue say:

"There are many persons to whom the assumption of an endless multitude of worlds appears difficult to reconcile with the belief of that which, as the Christian revelation teaches us, has been done for this our world of Earth."—P. xiv.

And, in his own person, he says:

"It may be useful if we can shew * * * that astronomy no more reveals to us extra-terrestrial moral agents, than religion reveals to us extra-terrestrial plans of divine government."—P. 133.

From this last quotation it seems obvious to conclude, that the writer's orthodox theology has, at least, *disposed* him to reject the idea of more worlds than one. His solution of the "astronomical objection against religion," as he calls it, is found by denying those vast conclusions which astronomers in general are disposed to draw from the discovered immensity and order of creation, and by maintaining for man (as he elsewhere does) "a nature and a place unique and incapable of repetition in the scheme of the universe."—P. 136.

His opponent, Sir D. Brewster, on the other hand, comes forth in his own name to maintain the majestic doctrine of "More Worlds than One," as the belief of the Philosopher and the faith of the Christian. But, as a member of the Scotch Kirk, equally strong in prescriptive orthodoxy with the most orthodox divine of the University of Cambridge, he too confesses, while he endeavours to solve, the "religious difficulty." He alludes, with very decided disapproval, to the theory of the late Dr. Chalmers in his *Astronomical Discourses*, as having "cut the knot of the difficulty rather than untied it," by "maintaining that the inhabitants of other worlds may not have required a Saviour;" and he thus deals with the difficulty in his own person:

"If we reject, then, the idea that the inhabitants of the planets do not require a Saviour, and maintain the more rational opinion that they stand in the same relation to their Maker as the inhabitants of the earth, we must seek for another solution of the difficulty which has embarrassed both the infidel and the Christian. How can we believe, says the timid Christian, that there can be inhabitants in the planets, when God had but one Son whom he could send to save them? If we can give a satisfactory answer to this question, it may destroy the objections of the infidel, while it relieves the Christian from his anxieties."

His answer to the question then follows:

"When, at the commencement of our era, the great sacrifice was made at Jerusalem, it was by the crucifixion of a man, or an angel, or a God. If our faith be that of the Arian, or the Socinian, the sceptical and the religious difficulty is at once removed; a man or an angel may be again provided as a ransom for the inhabitants of the planets. But, if we believe, with the Christian Church,* that the Son of God† was required for the expiation of sin, the difficulty presents itself in its most formidable shape."

Thus the orthodox astronomer expresses his own difficulty, in words which, ungracious as they are towards more rational Christian views than his own, need no exaggeration in order to make them truly grievous to every one who, possessed of a more comprehensive religious faith, sees the difficulty to be gratuitous, perverse and most unnatural. Let us then hear his attempted solution of it; and suppressing, if we can, the smile of contempt, yield rather the sigh of regret that such a scientific name as Brewster's should be thus coupled with an unscientific theology. He proceeds:

" When our Saviour died, the influence of his death

* Arians and Socinians, then, are not of the Christian Church!

† In the sense just defined, of being "a God who was crucified"!

extended backwards, in the past, to millions who never heard his name, and forwards in the future to millions who will never hear it. Though it radiated but from the holy city, it reached to the remotest lands, and affected every living race in the Old and in the New world." (A description, by the by, which would not be accepted by all orthodox people, and is quite inconsistent with the history of the influence of Christianity as actually spread by the apostles and first preachers in portions only of the Old world.*) "Distance in time and distance in place (he goes on) did not diminish its healing virtue:

> 'Though curious to compute,
> Archangels failed to cast the mighty sum!'

'Ungrasped by minds create,' it was a force which did not vary with any function of the distance. All-powerful over the thief on the Cross, in contact with its divine source, it was in succeeding ages equally powerful over the red Indian in the West and the wild Arab in the East. Their heavenly Father, by

---

* St. Paul says, "Whosoever shall call upon the name of the Lord shall be saved. How then shall they call on him in whom they have not believed? And how shall they believe in him of whom they have not heard? And how shall they hear without a preacher? And how shall they preach except they be sent?" (Rom. x. 13—15.)

some process of mercy which we understand not, communicated to them its saving power. Emanating from the middle planet of the system" (in what philosophical or conceivable sense the Earth is the middle planet of the system, he does not define), "why may it not have extended to them all? to all the planetary races in the past, when 'the day of their redemption had drawn nigh,' and to planetary races in the future, when 'their fulness of time shall come'?

'When stars and suns are dust beneath His throne,
A thousand worlds so bought were bought too dear.'"

Then, after all this magniloquence, he condescendingly proposes "to bring his argument more within the reach of an ordinary understanding." With this view, he makes an illustrative supposition to the effect, that our globe had been broken into two at the beginning of the Christian era, and that its two halves had thenceforth revolved together; in which case he concludes that "both its fragments would have shared in the beneficence of the cross" (the Theologian here again leaving the Philosopher unthought of, who might else have scrupled to admit the supposed survival of the human race through such a catastrophe); and he then asks whether, in like manner, "all the planets, *the*

*worlds made by our Saviour himself,*[*] formed
out of the same material element, and basking
under the same beneficent sun, may not be
equal participators in his heavenly gift?"—P.
138, &c.

Now who shall tell us, in the face of these
things, that theological opinions are of no im-
portance; that the day for scriptural controversy
is quite passed by; that the absurd dogmas of
former generations are dying out, and have only
to be let alone in order to be soon quite forgot-
ten? Would that it were so! But we are far
from such a point of progress as yet. Those
religious dogmas which are the poison of our
common human life and its social affections
among unscientific zealots, are the perpetual
stumbling-stone of science among the wise. Here
is a scientific astronomer, high among the highest
philosophers of our land, and possessing, moreover, a world-wide celebrity,—one of the eight
selected foreign Associate members of the Na-
tional Institute of France,—who yet writes like
a child, or a submissive servant of church creeds,

---

[*] Is this Sir D. Brewster's deliberate translation of
τὰς διῶνας ἐποίησεν (Heb. i. 2); or of καινὴ κλίσις (2 Cor.
v. 17, and Gal. vi. 15)?

in his attempted reconcilement of his religion with his philosophy. He confesses their apparent repugnance, and few intelligent persons will think that his proposed adjustment of their claims redounds much to the credit either of his theology or of his science.

The truth is, that the realms of scientific thought and those of the distinctive orthodox theology, so called, have *nothing in common*. Scientific men in general, for fear of open collision with the clergy, have agreed to keep their theology in abeyance; and then, without overtly calling in question the popular orthodoxy, they quietly pursue the great ideas of Science, and even imbibe its sublime religious thoughts, without bringing them into conscious juxtaposition with the theology of their churches. And so Revelation suffers from their hands one of two evils. It is either tacitly slighted or more actively doubted by the bulk of scientific men, who intuitively feel that there is no common groundwork of thought between the prevailing creeds of the churches and the revelations of science; while the few who venture to trace any mutual connections between natural science and what are currently assumed to be revealed truths, present such humiliating disclosures as are here before us. At the same time that the creeds of orthodox churches are thus virtually hostile to the free advance of science, the suppression of theological opinion and inquiry which the ascendancy of those creeds naturally dictates, is not less ruinous to religious simplicity and sincerity. Great need is there that those who so hold revealed truth as to present continually points of contact, and never of collision, with the truths of scientific investigation, should avow on the one hand their free and rational theology, and earnestly trace out on the other hand the great lessons of natural religion which science enables them more and more clearly to read. The separation of the two departments of Religion and Science is ruinous to both. Science, stopping short in the discovery of material substances, forces and laws, through fear of offending religious prejudices, becomes thereby the gross materialistic thing which it is often falsely charged with being in its own nature. And Religion, turning a deaf ear to the sublime discoveries of science, and keeping jealously aloof from contact with the intellectual activity of scientific thought, can never exert its highest inspiration upon the souls of those whose under-standings are closed in its sacred name.

And why is this separation? Why this jea-lousy between what is called Religion and what is called Science? Should they not be as parts of one and the same pursuit? the varied breath-ings of One Great Spirit? successive tributes of cumulative truth?—that truth having reference in both instances to the Creator's ways and will and attributes, and being properly religious when traced through the inductions of science, whether physical or mental or moral science,—and properly scientific, in the highest sense, when it has been caught from divine inspiration, in advance of man's natural knowledge of him-self, his duties and his destiny. Are they not God's works that the man of science explores, as truly as it is God's word that the divine, too exclusively perhaps, expounds? Is the language any more obsolete to our minds, in which the former are written, than the dead Hebrew and Greek of the latter, which are translatable, nevertheless, into living English? Or is their text more perplexed? Or are their versions more various? Or their readings more uncer-tain? Does not the Divine Spirit shine forth in both to the intelligent and religious-hearted reader? Is it not treason to both, to set them for a moment at variance? Does not the analogy which is traceable between Natural religion and Revealed, enrich the one and endear the other from their mutual sources?

So may the bright heavens ever shine upon the pages of the Gospel in the view of our free, comprehensive and blessed faith! So may the Gospel still shew our hearts a Heavenly Father's love in the mighty architecture of those countless worlds!

I ought perhaps to say a few words more specifically in reference to the two books to which I have adverted.

The book which holds the negative argument will strike most persons as decidedly the more clever of the two. It is lively, dashing and bold—may I venture also to say, one-sided and sophistical. The reply by Sir David Brewster is stately, rhetorical often, but heavy, and far less effective than those who hold essentially his side of the argument think it might and ought to have been. Nor is he free from that overstatement of his case, which most seriously of all enfeebles such a case as this, where analogy

is the ground of argument, and the analogy
grows fainter with distance till it is lost in doubt.
His way of speaking of his opponent, moreover,
is such as has been sometimes, but erroneously,
regarded as characteristic of theological controversy alone, and is certainly unworthy of Philosophy. Thus, he speaks of him as "exhibiting
an amount of knowledge so excessive as occa-
sionally to smother his reason." He ascribes
his sentiments "only to some morbid condition
of the mental powers, which feeds upon paradox and delights in doing violence to sentiments
deeply cherished, and to opinions universally
believed." He charges his opponent with em-
ploying "dialectics in which a large dose of
banter and ridicule is seasoned with a little con-
diment of science;" and describes his perform-
ance as "the most ingenious though shallow
piece of sophistry which we have encountered
in modern times." And he refers his "theories
and speculations to no better a feeling than a
love of notoriety." All this is in the worst
taste, if not the worst spirit too.

The opponent of a Plurality of Worlds rests
the strength of his argument upon the discoveries
of geology. It is a strange argument; but thus

he argues:—From the fact now (we may say) proved and accepted, that the Earth went through a variety of formations or states of growth, occupying long ages upon ages, before it became fit to be the abode of man, he argues (or rather he endeavours to reconcile the mind to the idea) that the whole Creation, this one world excepted, may be a similar waste for ever! He says:

"When Geology tells us that the earth, which has been the seat of human life for a few thousand years only, has been the seat of animal life for myriads, it may be millions, of years, she has a right to offer this as an answer to any difficulty which Astronomy, or the readers of astronomical books, may suggest, derived from the consideration that the earth, the seat of human life, is but one globe, of a few thousand miles in diameter, among millions of other globes, at distances millions of times as great."—P. 194.

To this Sir David Brewster has aptly replied, that if the comparison between time and space be allowed, the ratio of time in the habitable state of the Earth to time in its uninhabitable state is continually changing, and so vitiating the analogy to the assumed proportion between habitable and uninhabitable space. In fact, the analogy itself suggests the very opposite con-

clusion: If this Earth was for ages uninhabitable, yet in process of gradual preparation for
its present state,—then, if it could be proved
that other planetary bodies are now in its ancient geological state, it would be fair to argue
that they are in process of change and growth
towards a habitable state. This is the direct
argument from analogy.

The philosopher on the negative side seems,
indeed, to consider it possible that some of the
planets may contain animal life; but " a progressive creature, intellectually considered (he
insists) there is not the slightest ground for
believing to exist anywhere else" but on this
earth. (P. 121.)

Then he most wilfully argues that such a
superior being must be not only *like* man in
faculties, but *actually and identically* man;—
that " man cannot conceive any moral creature
who is not man" (p. 126), and so forth. Truly
a most wilful restriction of the analogical argument, which he permits, in the same sentence,
to fill some, at least, of the planets with plants
and animals, without insisting that they must
necessarily be *our* plants and *our* animals.

Then he as arbitrarily accepts the nebular

F

hypothesis for the Solar System, but stoutly denies it beyond (p. 314).

He calls "the Earth's orbit the Temperate zone of the Solar System" (p. 300); and the Earth itself its "domestic hearth" (p. 302);— his very figures of speech suggesting to us that Almighty Goodness may have formed appropriate inhabitants for the Torrid zone of Mercury and the Frigid of Uranus and Neptune (as for the varied zones of this planet Earth)—if rather the constitution of the respective planets themselves may not be the real adjuster and equalizer of temperature. And in the following bold metaphor he would reconcile us to the idea that this Earth is the sole ultimate purpose of the whole Creation :

"Instead of manufacturing a multitude of worlds on patterns more or less similar, He (the Creator) has been employed in one great work, which we cannot call imperfect, since it includes and suggests all that we can conceive of perfection. It may be, that all the other bodies which we can discover in the universe, shew the greatness of this work, and are rolled into forms of symmetry and order, into masses of light and splendour, by the vast whirl which the original creative energy imparted to the luminous

element out of which they were formed. The planets
and the stars are the lumps which have flown from
the potter's wheel of the Great Worker; the shred
coils which in the working sprang from his mighty
lathe; the sparks which darted from his awful anvil
when the Solar System lay incandescent thereon; the
curls of vapour which rose from the great caldron of
creation when its elements were separated. If even
these superfluous portions of the material are marked
with universal traces of regularity and order, this
shews that universal rules are his implements, and
that order is the first and universal Law of the heavenly work."—Pp. 353, 354.

Had this passage been found in any writer
between the days of Pythagoras and Copernicus,
or even if it could be conceived to have presented itself earlier still to a Hebrew mind, it
would have been deservedly regarded as suggesting a grand poetic thought, and doing homage
at once to the then natural idea of the comparative dignity of our Earth in the Creation, and
to the dimly-perceived reign of order beyond
the Earth. But, produced amid and after the
grand discoveries of size and distance with which
this age of the world has become acquainted,
the idea which it suggests is repudiated at once

by every smith at his anvil and every potter at his wheel, as known and felt by them to be unworthy of their own coarse manual art—how utterly, how degradingly unworthy of that grand Economy of Nature by which human artists and artificers have been hitherto accustomed reverentially to take copy! If the views here propounded be accepted, we must cease at least henceforth to cite the Economy of Nature as delighting to produce the greatest results by the simplest means, and as pursuing a grand and elevated utility even in the "pomp and prodigality" with which it ministers to our sense of beauty and our faculty of worship! No wonder that the writer of the reply should, in the severe spirit of retributory justice, have reproduced the following as the earlier and worthier thoughts of William Whewell in his *Bridgewater Treatise :*

"The earth, the globular body thus covered with life, is not the only globe in the universe. There are circling round our Sun six others, so far as we can judge, *perfectly analogous in their nature;* besides one moon and other bodies analogous to it. *No one* can resist the temptation to conjecture that these globes, some of them much larger than our own, are *not dead and barren;* that they are, like ours, occupied with organization, life, intelligence. To conjecture is all that we can do; yet even by the perception of such a possibility, our view of the kingdom of Nature is enlarged and elevated."—Brewster, p. 247; *Bridgewater Treatise*, pp. 269, 270.

And, in a corresponding tone, that Bridgewater Treatise speaks of the Fixed Stars:

"Astronomy teaches us, that the Stars which we see have no immediate relation to our system. The obvious supposition is, that they are of the nature and order of our Sun; the minuteness of their apparent magnitude agrees, on this supposition, with the enormous and almost inconceivable distances which, from all the measurements of astronomers, we are led to attribute to them. If then these are Suns, they may, like our Sun, have planets revolving round them; and these may, like our planet, be the seats of vegetable, animal and rational life; we may thus have in the universe worlds, no one knows how many, no one can guess how varied; but, however many, however varied, they are still but so many provinces of the same empire, subject to common rules, governed by a common power." — Brewster, p. 222; *Bridgewater Treatise*, B. iii. Chap. ii. p. 270.

Do we not prefer these first thoughts to the

second? May we not still rest in this truly grand and most devout philosophy, as sustained by the probabilities of analogical argument, reasoning cautiously and hypothetically, not mathematically nor dogmatically, from the great facts of Science?

## NOTE TO LECTURE III.

While preparing these Lectures for the press, I meet with the following estimate of the two books in question, in an able article in the *Christian Remembrancer* for Jan. 1855, which confirms the estimate given above. It also painfully illustrates what I have said of the hollow truce or transparent compromise effected, at Cambridge and elsewhere, between orthodox Theology and Science; for the reviewer, with an ingenious affectation of reverence, declines the solution of the religious difficulty, while ridiculing Brewster for his absurd attempt to solve it.

"In the one volume, a very bulky mass of very pompous thought, ingenious argument and elaborate dissertation, is expended in an effort to maintain a paradox and overthrow a popular belief. In the other, a highly discursive style of querulous remonstrance feebly advocates the established conviction, — sometimes by false philosophy, — sometimes by random theory, — sometimes by not indisputable analogy, — and not seldom by mere bombast."—P. 53.

Of Whewell's book the reviewer further says:

"Like all his other works, it evidences very great

faculties of mind, very comprehensive capacity of intellect, very extensive acquaintance with every branch of physical science; moreover, very clumsy and disorderly notions of logic, and a marvellous inelegance in the use of English."—P. 60.

Of Brewster's he speaks as follows:

"He takes up the cudgels against our Essayist, with a spirit worthy of better logic than he knows how to wield. The side he supports is the one we would advocate; but we blush for the forced analogies, the unsound argument, the daring speculation, the inflated diction, by which it is supported."—P. 78.

"We take leave of Sir D. Brewster's work, with a very acute feeling of disappointment. We are sorry to find so very famous a man condescend to employ rhodomontade and bad reasoning on so interesting a theme. We are the more sorry, because we feel that *such* an answer to so able (though we trust we have shewn not immaculate) a work as his opponent's, cannot but weaken the cause which we ourselves incline to support. To Sir David, an advocate of the Plurality of Worlds may well exclaim, 'Save me from my friends!' Would that the author of the Essay had been writing on the opposite side!"—P. 80.

# LECTURE IV.

SCIENTIFIC ANALOGIES AND THE CHRIS-
TIAN HOPE;,

OR,

OTHER WORLDS AND ANOTHER LIFE.

----

THE great religious impression produced on
our minds by Scientific Astronomy, is that of
the Immensity of Creation and the Infinite Perfections of the Creator.  Under its serene guidance, the world in which we dwell takes its
place as a part, and in some respects a subordinate part, of a system amid systems of worlds;
and from the known sizes and distances of objects upon the earth (some of which, indeed,
almost fill the practical limits of our clear perception of distance or of size), we are taught to
imagine, or at least to speak of, sizes and distances thousands and millions of times greater;

and then, again, we are required to multiply
the whole Solar System itself by an indefinite
number, till millions more or millions less can
make no real difference to our overloaded imagination :

> " The unsteady eye,
> Restless and dazzled, wanders unconfined
> O'er all this field of glories ;—spacious field,
> And worthy of the Master,—Him whose hand
> With hieroglyphics elder than of Nile
> Inscribed the mystic tablet, hung on high
> To public gaze, and said : Adore, O Man,
> The finger of thy God !"

Yet, however unable our minds may be to
realize the facts and their suggestions, it is
impossible for us to retreat from these dimly
sublime conclusions, because we have reached
them through steps of reasoning and calculation
which, in their earlier stages, are clear and definite enough, and which have only become less
clear and definite as their results have gradually
outgrown our power of grasping more.   Thus
Reason itself baffles Imagination; and lowly
Worship alone can express the mixed knowledge and emotion of the soul in the midst of
this discovered Universe.   In the presence of
such wonderful works, and of the mysterious

Being whose power, intelligence and goodness they proclaim, what indeed is Man, that the Lord of all those worlds is mindful of him; and the son of man, that He visiteth him with choicest blessings, making him little lower than the angels! In the presence of such a Universe, not only Man himself, his works and his interests, range themselves in due subordination and proportion in his own thoughts; but his religious ideas respecting Nature, Providence and Divine Revelation, all chasten themselves into order and harmony. He must cease to ascribe to the Divine Author of this wonderful Universe, any merely partial or local modes of operation, such as he might reverently have believed in while its majestic Laws were unknown to him. This Plurality of Worlds, while it rectifies all our paltry views of physical Philosophy, irresistibly bids us also expand and elevate our theories of Morals and Theology into harmony with that Universe of probable or possible intelligences. Natural Religion, thus dignified by the spirit of Science, forbids any low and narrow views of Christianity to seek acceptance thenceforth in her presence.

And while Astronomy, with its wonder-work-

ing Telescope, has thus *enlarged* our views of
Creation beyond our power of clear and definite
conception, the Microscope has carried out a
similar process in the opposite direction, shew-
ing us an extreme of minuteness not only beyond
our natural power of vision, but again beyond
our faculty of thought fully to realize, and
manifesting the same perfection of structure
and organization in the minutest forms which is
visible in the largest. And we find it equally
impossible to shew, or even to imagine, a limit
at either end of the series, beyond which Creation may not extend farther, or within which
its operations may not be still more minute,
than we have yet discovered or imagined.

Such mystery is in the Divine works! We
ourselves, a part of those works, to whom the
Creator hath given a spirit of understanding
which can discern His operations in them, are
soon perplexed and left behind in our attempt
to follow them on either hand, in the direction
of large or of small, beyond our daily common
experiences. But the limits of great and small,
it is evident, are in *our* faculty and *our* experience, not in God's creation. We are involuntarily making Him like ourselves, when we begin

to wonder how His creative power can reach wider and wider still, as our telescopes reach farther and farther yet without coming to the crystal bounds of space, or how it can contract itself to yet more and more minute organizations still, defying the research of our most improved microscopes. This is but our human point of view; for there is, in truth,

"To Him no high, no low; no great, no small;
He fills, He bounds, connects and equals all!"

In the discovered Plurality of Worlds, we trace the most instructive *analogies* prevailing through the remotest parts of the universe, and identifying all its worlds as subject to the same grand laws of One Sovereign Mind; while we also trace such *diversities* of *condition* resulting, as to impress us with the profoundest sense of the infinite diversity of *orders of being* which the Creator has, in His works best known to us, and must in those less known be conceived to have, endowed with varied life, intelligence and happiness.

In the two planets nearest to our own, the scientific astronomer finds reason to think that all the physical circumstances which are here connected with life and well-being, are *so nearly*

*repeated* as to warrant his belief that creatures
not much unlike ourselves and the other living
tenants of this Earth, may also inhabit them.
The amount of light and heat received is not so
much greater or less as to defy the thought of
adaptation in the creatures themselves, when we
bear in mind the adaptations of life to climate
between our own Equator and Poles. Even
the duration of day and night is almost identical
with ours. Atmospheres are seen to surround
the planets in question. The action of gravitation is not materially different in amount from
what we ourselves experience. And the conditions which cause our seasons are repeated almost
identically in one instance at any rate.* The

---

* Venus, compared with the Earth, receives $2\frac{1}{4}$ times
more light and heat upon an equal surface; Mars receives
somewhat less than one-half.

The revolution of Venus on her axis is computed at
something less than $23\frac{1}{4}$ hours; that of Mars is fully
ascertained to be something more than $24\frac{1}{2}$.

The action of gravitation on the surface of Venus is
almost identical with that experienced on the Earth; on
the surface of Mars it is less by one-half.

The inclination of the axis of Mars is 28 deg. 27 min.
(that of the Earth being $23\frac{1}{2}$ deg.); that of Venus is not
yet ascertained.

astronomer ventures to believe it possible that
man himself might live there, and that he would
find less difference in the physical circumstances
surrounding him, if he could be transferred to
that part of another planet which most nearly
resembles his own country here, than he often
does in passing from one extreme of climate to
another on this world of Earth. He ventures
to think it probable that such planets may even
be inhabited by beings organized much like
ourselves.

But when he passes on in thought to the
planets more remote in place from our Earth,
and differing more and more widely in these the
ordinary conditions of life as known to us; when
he finds one of them so much nearer to the Sun
that he can hardly avoid believing it intensely
hotter,* whatever modifying effect he may reasonably reserve for a particular constitution of
the body of the planet, or for the special atmosphere surrounding it; and when he finds others,
more and more remote from the Sun, whose
share of heat can seem but small comparatively,*

* Mercury being three times nearer to the Sun (in
round numbers), receives nine times as much light and
heat on a given surface as we do; Jupiter, on the other

however modified it may reasonably be supposed
to be by the nature of the material composing
such planets, and also by that of their respective
atmospheres, and whose diminished share of
light is evidently compensated to a small degree,
but in no sense equalized, by their attendant
moons; when he also finds reason to conclude
that these remoter planets are composed of mate-
rials very much lighter than the substance of
the Earth (the specific gravity of Jupiter, for
instance, being compared to that of the heavier
kinds of wood, Saturn to lighter wood, and
Uranus and Neptune being apparently about
the weight of water);—his ideas of the inhabitants that might be suitable for these planets
respectively, must be greatly modified and more
and more vague continually. Here, while the
grand analogy is unbroken, inasmuch as gravitation, light and heat act by universal laws, the
intensity varying in a known ratio to distance;
that varied distance itself changes the result so
greatly, as to forbid us to adhere too closely to
the definite forms of life with which we are

hand, receives only the $\frac{1}{25}$th part, Saturn $\frac{1}{81}$, Uranus $\frac{1}{324}$, and
Neptune $\frac{1}{784}$, being respectively at 5, 9, 18 and 28 times
our distance from the Sun.

familiar on this Earth, in our reverent attempt
to people those distant orbs. But then the differences of circumstance, and corresponding differences of organization, under which we see
the life-giving agency of the Creator manifested
on this Earth, are sufficiently wide, surely, to
suggest to the scientific imagination wider diversities of formation still, corresponding to yet
wider differences of circumstances. Surely we
do not doubt the possibility of countless other
forms of life, besides those with which we are
actually acquainted! Who that had never seen
a fish, would easily conceive that living creatures
could exist and breathe immersed in water?
Who that had never seen a bird or a winged
insect, could readily conceive how creatures
might be formed capable of passing through the
thin air without touching the ground? And
when we trace the infinite diversities of living
creatures that inhabit this globe, endlessly varying with climate and locality, but everywhere
adapted to the position assigned them in the
great Economy of Nature, the profoundest as
well as the most religious reflection that suggests
itself surely is this,—that the Creator has forms
of life suitable to every imaginable difference of

condition. Our Earth itself not only exhibits such varieties through land and sea and air, from the Torrid zone to the Frozen poles; but also, in the fossil remains of extinct species, preserves the types of creatures suitable to an earlier period of our own planet, before it became fit for human occupation. And if the scientific naturalist admiringly traces all these endless varieties of creatures to a few archetypal forms, evincing the severe simplicity of Nature's plan, he should beware of so unscientific an assumption as that of believing that there can be no other types of life in other worlds, besides those which he sees in this. Rather should these seen varieties suggest to him the largest general faith in Creative skill and goodness, and assist him vaguely to imagine that, however different the physical circumstances of other planetary bodies may be from those which prevail in this globe, there may be other orders of creatures, living and rejoicing, and with intelligent beings at their head, all equally appropriate in their physical structure to the different physical circumstances of the worlds in which they respectively live.

And Man himself, the lord of all below,— how various both in physical and spiritual constitution, circumstances and developement, is he! Not another creature in this world is found existing under such various circumstances of climate and other physical influences, or exhibiting such varieties of bodily and mental character, as Man. If one human race, presiding over this planet, is thus diversified in the external forms of Papuan, Negro, Hottentot, Red Indian, Asiatic and European,—and in its mental developements from the hunter in the wilds to the civilized worker and thinker, from the savage idolater to the philosophical Christian,— need we doubt the possibility of other orders of being, analogous in position, yet for that reason diverse in formation from ourselves (though we know not *how* formed), to whom the supremacy may have been respectively assigned in every other habitable world that God hath made?

Such are among the suggestions of the great argument from Analogy, implying endless diversities of operation on the part of the One God, who is in all and through all and above us all. It shews us, actually existing, so much that is wonderful, majestic and benevolent, as to enable us to believe that whatever else is required

throughout the universe, of wise and wonderful
and benevolent, is everywhere done.

There is a most weighty and most interesting
topic of religious belief and religious affection,
which more particularly connects itself with
these speculations on the Plurality of Worlds.

Involuntarily and rightly, the thought of the
FUTURE LIFE seeks to find for itself a resting-
place among those dimly discovered worlds, as

" Perhaps our future home, from whence the Soul,
    Revolving periods past, may oft look back
    With recollected tenderness on all
    The various busy scenes she left below,
    Its deep-laid projects and its strange events,
    As on some fond and doting tale that soothed
    Her infant hours."

Does our Astro-theology point out the probable, or possible, locality of the future life of
man ?   Or (short of this) does it help to support or define, in any way, that natural hope
and Christian belief ?

The true science of the heavenly bodies has
long since dispossessed the popular Heaven and
Hell of their prescriptive seats above the blue
sky and below the dark earth.   Our antipodes

(as everybody knows) dwell, like ourselves, upon the opposite surface of the round earth, where the old Greeks, in ignorance of its globular form, placed their Tartarus and Elysian fields. Within the earth, the Geologist cannot find fit space even for the Roman Catholic Limbo, or Purgatory. And the Heaven of ancient Jewish and Gentile belief above the blue firmament, has dissolved into space occupied, at intervals, by suns and systems of worlds. Has Science, then, while thus disproving these definite false beliefs of former ages, given a definite true one in their place? Has it shewn us *where*, or *how*, or *when* that Future Life is?

It has not. It has not told us definitely, what is or will be. But indefinitely, what *may* be, in hundreds and thousands of ways at the Sovereign Will, it has told us and proved to us. It has disclosed orderly worlds so numerous and so vast, reasonably believed to be inhabited or habitable, as to make it an absurdity for any philosophical believer in human immortality to ask doubtingly, *Where* the future life of man *can be?* Can be? If that is the question,—it can be *here*, if God hath so willed. A spiritual life may be well conceived of as tenanting the

self-same abodes in which it erewhile dwelt in
the flesh, now unseen by us who remain, though
taking earnest and loving interest in us.    And
here, in the very scenes of its fleshly dwelling,
the spirit may be reaping the blessedness or the
woe of its previous actions and character, with
sharpened perception of the one unalloyed by
earth's trials, and of the other no longer dis-
guised by outward indulgence.    *Can be?*    It
*can be* in any planet in the Solar System, if God
hath so willed ;—in one of those nearest and
most like the earth in their physical circumstances, if it hath pleased God to ordain that,
in the life to come, we shall be reinvested with
bodily limbs and senses closely resembling our
present lot ;—or it *can be* in any of those least
like our own, if it please God to invest our
souls with proportionately different organism
(we know not how different, nor what it should
be like), adapted to we know not how different
physical circumstances!    The question of *Can
be*, is answered from planet to planet through
our system : *Here, here! There, there! if but
the Sovereign Will commands.*    And then, from
star to star along the boundless heavens, if
the restless soul would travel farther in search

of its possible future, the vaguer answer is still harmoniously echoed further and further back from boundless space: *Where can it not be? Where can it not, if the Almighty Father give the word?*

But, if the question be, Where *is* the future life of man? or, Where *will* it be?* religious Philosophy is not ashamed to say, I know not, nor need I know, since it is, or it will be where the good God chooseth.

Certain difficulties, supposed to be derived from the New Testament Scriptures (but not, I believe, properly chargeable upon them), are commonly felt, indeed, by religious persons both in conceiving the general doctrine of a Future Life, and in connecting it specifically with these scientific views of the starry worlds.

" The resurrection of the body" is one of the articles of the creed of the Church of England: " I believe in the resurrection of the body and the life everlasting." Now, to many minds, this

---

* The question between *is* and *will be*, that is between immediate and deferred restoration to life, is a question of consciousness and mental philosophy, rather than of Scripture or Astronomy. The latter can only repeat, that unnumbered worlds exist.

juxtaposition of terms seems little better than a direct contradiction. The bodily life which we now live is, through the divinely-appointed laws of its natural constitution, temporary only; and a life everlasting seems necessarily to imply a very different state from this bodily one. And how, again, can the astronomer imagine bodily transference from this earth to another planet or another system?

If we ask those who recite this creed, what they understand by believing in the resurrection of the body, they will generally declare that they believe the identical body which has died will be raised again by the power of God and made immortal, like Christ's glorified body in which he rose from the dead. They think this is the doctrine of Scripture; and they try to believe it, without asking whether it is credible or not. But Philosophy puts in a serious doubt, and cautions us not too rashly to ascribe to Christianity a doctrine which may prove to be not true. It asks us how it is possible that the same mortal body which has died and been decomposed and scattered, it may be, to the winds, or which (even in the most common case, of Christian burial) has entered into new material combinations, forming in part the substance of plants and then of animals, and thus partly entering, no doubt, into other human bodies, can be reconstructed of the very same particles which belonged to it before death. Reverently we feel, that Deity itself cannot do inconsistencies. Devoutly we believe, that the Author of the Laws of Nature respects His own enactments. Chemistry declares, in the name of those Divine Laws, that the thing under consideration is impossible. Chemistry tells us, that even during the course of a few years of life, every particle that forms our body is changed and replaced by others. How needless, then, to assert a *bodily* identity after death, which is lost even during life! Not Divine power can be believed to effect things divinely constituted inconsistent. The particles which have belonged to a certain human body at death, and which have since gone into the composition of various plants, animals and other human bodies, cannot be collected together, identically the same particles in each instance, to form the substance of two or three different bodies in the resurrection. Seriously to maintain such a doctrine is only to bring religion into contempt. No philosophical mind can believe in the re-union of identically the same bodily particles which form the body before death, at a future resurrection of that body to eternal life. And in the name of Philosophy we must hope that Revealed Religion is not responsible for any such doctrine.

Why, indeed, we may well ask, should the Church make it an article of faith? It is nowhere so declared in Scripture. I say, distinctly and very deliberately, I do not find this doctrine of the Resurrection of the Body to be a scriptural doctrine at all.

It is indeed a scriptural fact, quite beyond all doubt, that *Jesus Christ*, having died, rose again, *without having seen corruption*, the same bodily man that he died. But that the human body of Christ lives eternally in a heavenly state may well be doubted; and (which is the real point to be noticed) the case of his resurrection from the dead is miraculous and exceptional, and bears no analogy of manner or form to ours. He " saw not corruption;"—we do. His mortal body rose to human life again the third day;— ours do not. Therefore, the fact of Christ's special resurrection from the dead in the same body in which he died, is far indeed from bearing out the inference that we too shall rise from the dead in the identical bodies in which we die. The comparison does not hold good in any essential point. The one case is no true precedent for the other. Nor is any such overstrained analogy necessary for the Christian faith in Immortality. *Identity of soul* is all that the doctrine of the future life requires; and the soul sees no corruption in the body's mortality.

St. Paul is indeed thought to favour this doctrine when, in that well-known chapter of his first Epistle to the Corinthians which forms part of the Burial Service, he deals with the objections of those who, doubting or denying a future life, asked tauntingly, " *How* are the dead raised up, and with *what body* do they come ?" Now a careful reader of that chapter will, I think, soon perceive that Paul does not attempt to give a categorical and dogmatic answer to the question, With what body do they come ? He only says in effect: God will give a body as He pleaseth; and man may be satisfied to rely upon the power of Him who has promised that the dead shall live.

In illustration of the reasonableness of this trust, he refers the doubter to one of the com-

monest phenomena of nature,—the growth of
the corn which he sows in his field,—to shew
him the absurdity of his sceptical *how*. He
cannot tell how that corn grows; yet he believes
it does, for he sees it. If he saw the "bare
grain" for the first time, without ever having
seen the plant, he could not possibly tell with
what body it would come up. He could not
possibly anticipate "first the blade, then the
stalk, then the ear, and then the full corn in
the ear." But *God giveth it a body, as it hath
pleased Him.* Here is a genuine and deep mystery in one of the commonest occurrences of
nature. Then, again, every seed has its own
body. Each different kind produces a different
kind of plant. Who could anticipate, from
looking at the different seeds, the different
plants that will grow from them? From strange
seeds imported from a distant country, who can
conjecture the new varieties of stem and leaf
and flower? Or what can all our Physiology
and Chemistry even yet say about the processes
of vegetable life and developement? Absolutely
nothing, as to *how* that body is given to the
plant in which it rises to light and beauty before
God and man. "So," says Paul, "is also the

resurrection of the dead,"—equally inexplicable, but equally certain. *God will give a body* there. The seed of mortality will become an immortal growth. We know not how, but we know that God hath boundless power and endless diversities of operation.

Guided by his rich imagination and fervid feeling, the apostle finds further illustrations, popular rather than scientific in form, and yet in their spirit perfectly scientific and profoundly philosophical, when he turns his glance to the heavenly bodies. To him those bright orbs shone with the poetry of true devotion. "There are bodies celestial," he says, "and bodies terrestrial; but the glory of the celestial is one, and the glory of the terrestrial is another. There is one glory of the sun, and another glory of the moon, and another glory of the stars; even one star differeth from another star in glory. *So also* is the resurrection of the dead,"—the difference as striking, and equally appropriate the change,—"from corruption to incorruption, from dishonour to glory, from weakness to power, from an animal body to a spiritual body." And with this most expressive figure, amounting to a verbal contradiction deliberately and purposely committed,—the "spiritual body,"—he completes the climax of his thought and feeling, at once rebuking the poor materialistic doubt and instilling a spiritual faith.

Do we not feel, in our own consciousness, that the future life to which it is at once the deepest necessity and the proudest privilege of our nature to cling, is the life of the *Soul?* It is not for our fleshly limbs and senses that we crave immortality; but for our powers of thought and affection, with whatever body they may be endowed, or whether with a body (properly speaking) or not. Even in this life it is often proved to us, by experiences at once sad and holy, that the bodily limbs or organs, with their specific infirmities or defects, are not the conscious Man himself; for in the sad loss of one or more, not affecting the vital functions, his conscious self remains. Even in this life it is often proved that the bodily senses which we possess are not, in their precise number,—no more, no fewer,—required in order to constitute a human being with spiritual faculties ready for developement; when in the deprivation of one or other of these precious inlets of knowledge and impression, the thinking and feeling soul may still be developed, though with more difficulty to the sufferer, pleading for most special care and help on the part of others. Thus its present being is not absolutely dependent upon each and all the usual limbs, organs and senses of the human frame. Who then shall tell us what different, what more numerous, what more perfect inlets to knowledge and goodness may be given to man when made immortal, or which of his earlier implements of knowledge and action may be superseded?

These things it is not given to us either to know or to conceive in actual detail, while in our present state. A child cannot imagine the faculties, wants, feelings, passions, aspirations of manhood; nor may the mortal imagine those of his immortal state. We can only gain that large faith in the infinite resources of Divine power, wisdom and love, which makes the more minute answer quite unnecessary for our peaceful and sublime trust.

This faith, then, we cultivate and sustain, when in the spirit of Christianity we study the exalted science of the star-lit heavens. In this way only, and only to this extent,—but assuredly to this important extent and in this most effectual manner,—does the scientific idea of other Worlds sustain and inspire with philosophic argument the Christian promise of a Life to come.   And the philosophical thought, like the Christian promise, lights us back wiser and happier to the home scenes of present duty, trial and love, to leave upon them the inspiration of great thoughts and immortal affections.

" For soon the soul, unused to stretch her powers
    In flight so daring, drops her weary wing,
    And seeks again the known accustomed spot,
    Drest up with sun and shade, and lawns and streams,
    A mansion fair and spacious for its guest,
    And full replete with wonders.   Let us here,
    Content and grateful, wait the appointed time
    And ripen for the skies.   The hour will come
    When all these splendours bursting on our sight
    Shall stand unveiled, and to our ravished sense
    Unlock the glories of the world unknown."
                                        BARBAULD.

C. Green, Printer, Hackney.

Lightning Source UK Ltd.
Milton Keynes UK
UKHW020647070223
416609UK00011B/2470